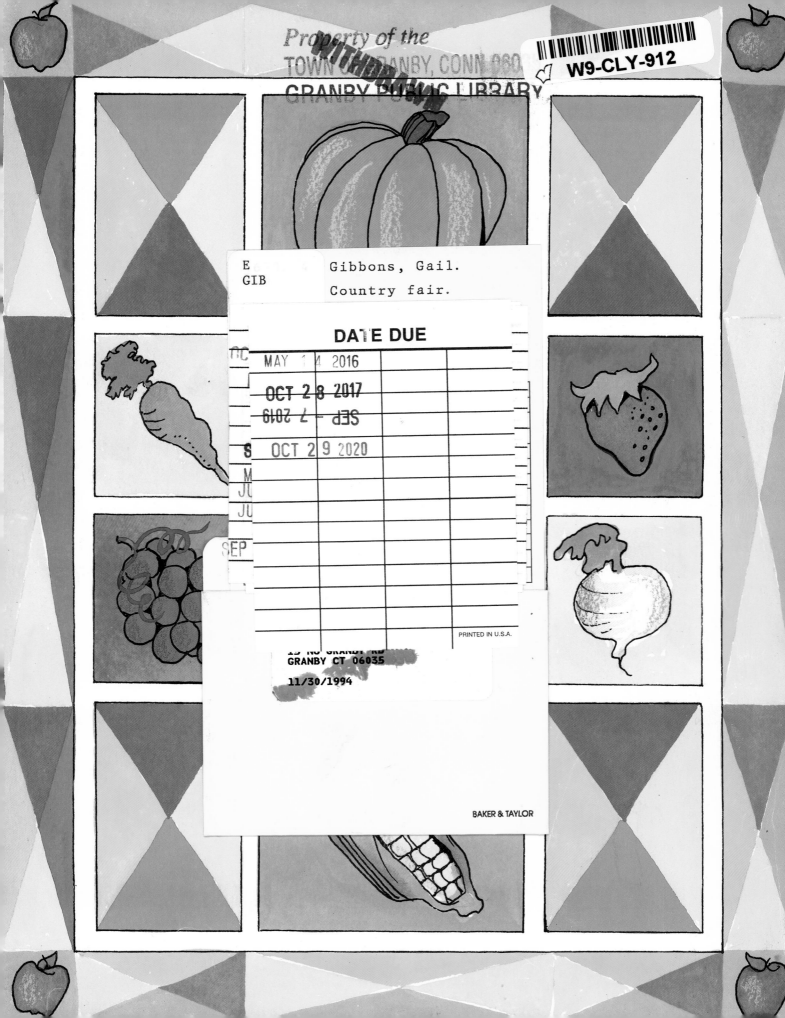

COUNTRY FAIR

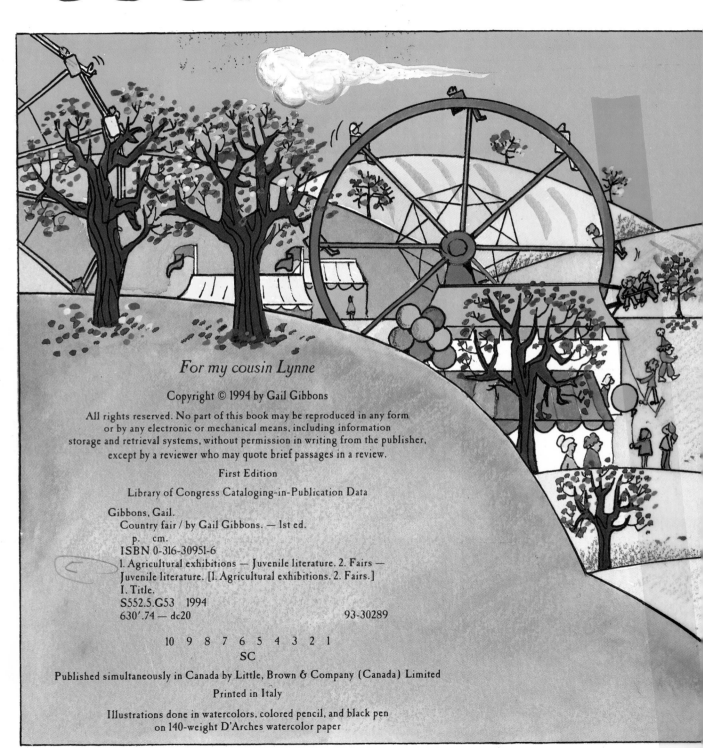

For my cousin Lynne

First Edition

Library of Congress Cataloging-in-Publication Data

Gibbons, Gail.
    Country fair / by Gail Gibbons. — 1st ed.
    p.   cm.
    ISBN 0-316-30951-6
    1. Agricultural exhibitions — Juvenile literature. 2. Fairs —
Juvenile literature. [1. Agricultural exhibitions. 2. Fairs.]
I. Title.
S552.5.G53   1994
630'.74 — dc20   93-30289

10 9 8 7 6 5 4 3 2 1
SC

Published simultaneously in Canada by Little, Brown & Company (Canada) Limited

Printed in Italy

Illustrations done in watercolors, colored pencil, and black pen
on 140-weight D'Arches watercolor paper

by Gail Gibbons

Little, Brown and Company

BOSTON   NEW YORK   TORONTO   LONDON

It's a crisp autumn day. Everyone is busy in the valley. Carpenters nail stands together. Workers tidy up buildings, mow the fields, and sweep the tracks.

Some exhibitors assemble their booths. Others lead farm animals to their stalls. Rides and games are put into place. Everyone gets ready for the next four days.

A new day begins. Hooray! It's opening day for the country fair! Cars and trucks begin to fill a farmer's field.

People from all around head for the admission gate to buy their tickets.
The high school marching band leads the way.

"Buy your cotton candy here!" a worker cries out as people enter the fairgrounds. *Oink...baa...moo...* Animal sounds fill the air.

Families and friends stroll down the midway looking at everything. Tents and stands fill both sides of this main route down the center of the fairground. There are games to play and much to see. "Test your skills!" a man yells from the ringtoss booth.

Clang! A very strong person just made the bell ring. Nearby, a bunch of kids throw balls at targets. Outside the bingo tent, a clown paints and decorates children's faces.

Refreshment stands full of popcorn, french fries, pizza, and fried dough are bustling with activity. Chicken sizzles on the grill, and there's a long line at the hamburger stand. "We make our own ice cream," a woman tells her customer proudly.

The Ferris wheel slowly turns, taking its passengers high up into the sky. A balloon floats by. What a view there is at the top! The merry-go-round plays a lively tune as the riders go up and down...up and down.

A line snakes in front of the Mad Mouse ride. Passengers on the ride shriek as they race around the track. Everyone loves the rides at the country fair!

Scary sounds escape from the entrance of the Haunted House. Is that a skeleton laughing? A magician in a top hat dazzles the audience with her tricks. A clown carrying a bunch of balloons walks by on stilts.

Many country fairs celebrate the fall harvest. Inside the biggest exhibit hall, people display prized fruits and vegetables from their summer gardens. A row of pumpkins lines one wall, and tomatoes, squash, green beans, and other vegetables are on display as well. The best entry in each category is awarded a blue ribbon.

Home-canned vegetables and fruit line the shelves. There are jellies and jams, and maple syrup will be judged, too. People show off pies, cakes, muffins, and loaves of bread for prizes as well.

Flower arrangements look beautiful and smell like a real garden. A judge strolls by, deciding which rose display will win first prize. The tallest sunflower already has a blue ribbon attached to it.

Arts and crafts are on display in tents outside the exhibit halls. Paintings, sculptures, and photographs are being exhibited in one tent. And next door, one can see handmade crafts such as needlework, pillows, pottery, and quilts. "My quilt won a blue ribbon!" one girl says to her friend.

Outside, people look at the latest in farm equipment. "Have I got a deal for you!" one salesperson says to his customer.

Over at a fenced-in area, a calf judging is taking place. One by one, the calves parade through the center of the arena as the judges take notes. Some of these calves are owned by members of the Future Farmers of America or the 4-H club. They've brushed their calves' coats until they're bright and shiny. "The first-prize winner is..."

Other farm animals rest in their stalls, waiting their turn. The oxen are huge!
One girl feeds her goats as sheep bleat, chickens cackle, and pigs squeal.

Not too far away, in another exhibit building, cows and calves moo, while
outside, a pig-calling contest is going on. "Sooo-eeee!" cries a little boy.
Everyone cheers.

Spectators line the fence of a big arena. Riders show their horses as they trot around and around. Some of the horses have braided manes and wear fancy ribbons. Judges stand in clusters, taking notes and awarding prizes.

At the stalls, the riders feed and water their horses. They brush them down and get them ready for the next event.

Next door is a model of a one-room schoolhouse. A little girl in overalls stands outside ringing the bell. Inside, a woman dressed up as an old-fashioned schoolteacher writes on a chalkboard and answers questions.

A fiddlers' contest is taking place outside the schoolhouse. Across the way, folks young and old demonstrate contra dancing. A man shouts out dance calls as the dancers swing around. The music plays, and everyone claps along to the beat.

Nearby, in a roped-off area, antique buggies, sleds, and tools are on display. Some folks working at the fair put on demonstrations of how things were done in the olden days. A blacksmith hammers a hook into shape at his anvil. A woman pours hot melted wax into old-fashioned candle molds. This batch is lavender scented. Some people spin wool and use their homemade yarns for weaving. There's even an antique cider press in full operation. "Come and get it!" calls a barker.

At the big grandstand, people cheer on the drivers in a horse-and-buggy race. Look at them go!

Off in the distance, people polish their tractors and rev the engines. They are getting ready for the next big event, the tractor pull. Which one will pull the most weight?

As evening comes to the fair, the grandstand's lights flood the arena with brightness. A band plays country music. The seats fill up and everyone — young and old — sways to the music.

Now it is nighttime. The moon sits high in the sky. Fireworks light up the darkness, bursting into spectacular patterns. *Bang! Pop!* It's a perfect way to end a perfect day.

PLANNING A COUNTRY FAIR

It takes a lot of time and many people to put a country fair together.

First, committees are formed to do different jobs. During the winter, spring, and summer months, they hold meetings. They huddle over their papers, planning the event. They schedule entertainers and arrange for rides, games, and other amusements. They choose judges.

As opening day comes closer, posters, programs, and maps are printed. Then advertising begins. Posters go on display. Ads are run in local newspapers and on radio and television.

It's time for another country fair!

A HISTORY OF FAIRS

The word *fair* comes from the Latin word *feria,* which means festival.

In ancient times, thousands of years ago, people lived in settlements that were far apart. They came together from these places to worship their gods at festivals. Often people sold their goods. This is how fairs began.

During the Middle Ages, between the years 500 and 1500, many fairs were held throughout Europe. Often, these fairs were held in honor of Christian saints. They became centers of trade, too.

Around 1200, in England, the Bartholomew Fair began. It was the first fair to have circus acts and amusements.

In 1644, the first country fair in the United States took place in New Haven, Connecticut. This fair showcased farm animals and products.

In 1810, the first Berkshire Cattle Show was held in Pittsfield, Massachusetts. For the first time, prizes were given for handmade crafts, cooked goods, and farm animals.

Today, country fairs have grown in importance. More than two thousand are held in the United States each year. Millions of people go to them.